A Teacher's Guide to
A Drop Around the World

Lesson plans for the book
A Drop Around the World, by Barbara Shaw McKinney

by Bruce & Carol Malnor

Bruce and Carol Malnor together have over 40 years of educational experience. Bruce has been a classroom teacher, elementary school principal, and is a Brain Gym instructor. Carol has taught elementary, junior high and high school, and has helped found two alternative high schools. They are directors of the Education for Life Foundation and have conducted workshops for educators throughout the U.S., as well as in Canada, Germany and Italy.

*Sharing Nature
with Children Series*

Dedication

To our parents and all of the teachers
who have helped and inspired us.

And to all teachers around the world.

LET THE CHILDREN TAKE CARE OF THE EARTH,
CONSERVE IT, PROTECT IT, AND VALUE ITS WORTH.

Publisher's Cataloguing-in-Publication
(Provided by Quality Books, Inc.)

Malnor, Bruce.
 Teacher's guide to A drop around the world / by Bruce and Carol Malnor.
 p. cm. — (Sharing Nature with Children series)
 Includes bibliographical references.
 ISBN: 1-883220-77-7

 1. McKinney, Barbara Shaw. Drop around the world. 2. Hydrologic cycle—Study and teaching (Elementary)—Activity programs. 3. Raindrops. 4. Teaching—Aids and devices. I. Malnor, Carol. II. McKinney, Barbara Shaw. Drop around the world. III. Title.

GB848.M35 1998   577.48'071
                                  QBI98-237

Published by DAWN Publications
12402 Bitney Springs Road
Nevada City, CA 95959
800-545-7475
e-mail: nature@DawnPub.com
website: www.DawnPub.com

Printed in Canada

10 9 8 7 6 5 4 3
First Edition

Illustrations by Sarah Brink
Computer production by Rob Froelick and Renee Glenn
Designed by Lee Ann Brook

Table of Contents

Dear Teacher,

A single drop of water is magical! It is a drop of life! It touches all plants, animals, and humans as it travels around the world. A single drop of water can deeply touch the lives of your students in meaningful ways. By using *A Drop Around the World*, you can help your students explore the beauty and wonder of our natural world while also helping them discover the beauty within themselves. As educators, we have the responsibility to open our students' hearts and minds to the world around them; the world for which they will be responsible. Tanako Shozo, a Japanese conservationist, observed, "The care of rivers, is not a question of rivers, but of the human heart." It is essential for students to not only intellectually understand science concepts, but also to develop skills that will help them be compassionate stewards of our world. *A Drop Around the World* combines science with poetry, core curriculum with character education, the head and the heart.

This guide for *A Drop Around the World* contains three components: **Water Magic**—seven lesson plans about the water cycle, properties of water, and role of water on the Earth and in our lives; **World Habitats**—background information and maps on seven habitats; **Skills for Living**—eight lesson plans about character qualities that water teaches. The lesson plans are designed so that they can be used individually or as part of a six-week unit. The lesson plans are most suitable for grades three through six, but can be adapted for some seventh or even eighth grade classes. Unless otherwise stated, the lessons can be completed in a 45-60 minute class period; a few may need additional time. This time can be adjusted by including more or less student discussion and sharing. For teachers who use block scheduling, many lessons can be combined or used back to back, creating a full 90-120 minute class session.

Each lesson includes the following elements, which in our experience provide the greatest possible educational impact: *Flow Learning™ Format; Tools of Maturity; Benchmarks; Skills for Living; Mind Mapping; and Brain-Compatible Activities.* These components are explained more fully on the following pages. A **Six-Week Unit Plan** which includes suggestions for additional student projects is also described.

We invite you and your students to join "Drop" in an exciting learning adventure.

Bruce and Carol Malnor

The Whys and Hows

Flow Learning™

Why: Simple tools can be some of the most effective. Flow Learning™ is one such tool for us as educators. It was developed by Joseph Cornell, the famous nature educator and author of *Sharing Nature With Children*. We use Flow Learning™ because it is based on how people learn. The Flow Learning™ process captures students' interest in the lesson right from the beginning, thus eliminating or minimizing many discipline problems. It's a simple, effective way to uplift a group's energy. There are four steps in the Flow Learning™ process:

1. Awaken Enthusiasm—Children learn if the subject matter is meaningful, useful, fun, or in some way engages their emotions. Time spent in creating an atmosphere of curiosity, amusement, or personal interest is invaluable because once students' enthusiasm is engaged, their energy can be focused on the upcoming lesson.

2. Focus Attention—Some students' minds can be compared to a team of wild horses running out of control. Without concentration no true learning can take place. The power of a laser beam lies in its intense focus; so it is with our thoughts. This guide uses the engaging picture book, *A Drop Around the World*, by Barbara Shaw McKinney to focus attention and increase learning capacity.

3. Direct Experience—Once students' interest and energy is awakened and focused, the stage is set for a direct experience. Each lesson plan is designed to provide an experience that expands the students' knowledge base, provides an opportunity to use the information to create or synthesize something new, or inspires new awareness.

4. Share Inspiration—Each lesson provides an interesting way for students to reflect together on what they have learned. In our fast-paced world, students and teachers alike often rush from one activity to another. Yet taking the time to reflect upon an experience can strengthen and deepen that experience. It need not take long. It can be as simple as responding to a few questions, writing a journal entry, or drawing a picture. Goethe said, "A joy shared is a joy doubled." Giving students the opportunity to share their experience increases the learning for the entire class.

How: You can experience and understand the Flow Learning™ process by simply following the lesson plans. As its name suggests, Flow Learning™ is flexible, so feel free to make adjustments in the lessons. Some classes may need greater emphasis given to Awaken Enthusiasm activities while others may need more time for Focus Attention. Shorten or lengthen these parts of the lesson to suit your students' specific needs. The Direct Experience meets the stated objective of the lesson; choose activities according to what you want your students to learn or experience.

Benchmarks

Why: Maintaining high standards in the classroom aids student achievement. The benchmarks in *Content Knowledge: A Compendium of Standards and Benchmarks for K-12 Education*, by John S. Kendall and Robert J. Marzano, identify the skills and knowledge which are essential for all students.

How: Each lesson plan in this guide identifies one or more benchmarks which primarily relate to Science, Life Skills or Language Arts. Choose activities which will meet the standards your students are working towards.

Skills for Living

Why: One of our goals as teachers is to help our students lead successful lives. True success is measured not by material standards, but by quality of life: happiness, fulfillment, joy. The attitudes and qualities which lead us toward true success are called Skills for Living. Like any skill, they can be taught and practiced. Students' character development affects not only them, but also the future of our country and our planet.

How: Each lesson focuses on one or more of the Skills for Living. Simply doing the activities gives students practice with the skill. It can also be helpful for students to have these skills identified by name so that they can understand the skill more completely and apply it in other contexts. With continued practice and application, students internalize the Skills for Living and make them their own. A list of Skills for Living can be found on page 41.

Mind Mapping

Why: Graphic organizers have many different names: story webs, semantic maps, concept maps, and Mind Maps. We use the term "Mind Map." Mind Maps are visible representations of information using either pictures, words, or a combination of both. Through Mind Mapping students develop organizational skills as well as thinking skills—both the sequential and gestalt (left and right) hemispheres of the brain are used. Comprehension and memory increase as students Mind Map information they read or hear.

How: Mind Maps can help students organize their research information. The following illustrates how to Mind Map some of the information about the Desert habitat. Mind Mapping is taught more completly in the Sharing Nature With Children Teacher Guides for *Lifetimes*, *A Walk Through the Rainforest*, *A Swim Through the Sea*, and *A Fly in the Sky*.

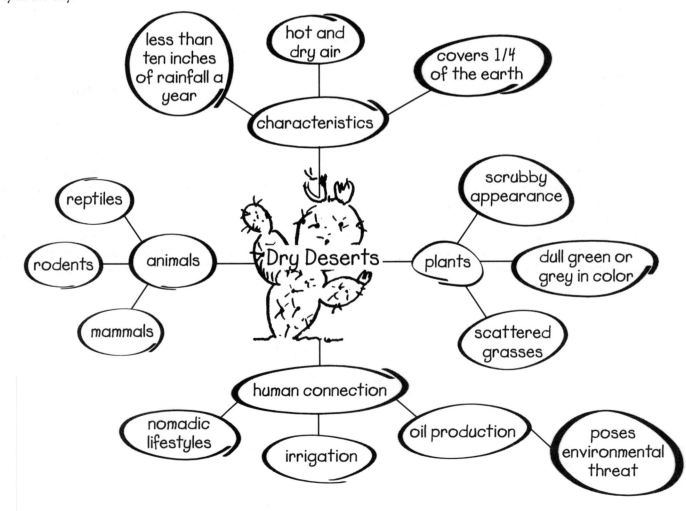

Brainstorming, Discussions, and Sharing

Why: Brainstorming, discussions, and sharing provide students with various avenues of self-expression. The type of student grouping determines the outcome:

- With a partner—Students sharing one-on-one gives everyone a chance to talk without taking a lot of class time.
- In a small group—Collaboration stimulates creative thinking.
- Presented to the whole class by individual students—Students have opportunities to practice poise in front of a group.
- Privately through a journal entry—Non-verbal activities balance the usual outward way of sharing with a more introspective approach.

We encourage you to experiment with different types of student groupings to suit the particular dynamics of your class. Variety helps to keep student responses fresh and enthusiastic.

How: It is essential to establish an environment in which students can share without fear of ridicule. Before beginning any brainstorming activity, review the guidelines for brainstorming with students: (1) Every response is acceptable; tell students not to be concerned with accuracy at this point in the lesson. (2) Encourage lots of variety and unique responses; remind students that the "craziest" response often leads to the best idea. (3) Absolutely no put downs are allowed; have clear consequences for any student who uses a put down.

The conclusion of a lesson is the time to acknowledge and celebrate student learning. Teach students to share their thoughts and feelings by (1) sharing your own thoughts and feelings, (2) giving heart-felt appreciative comments to students, (3) asking reflective questions, (4) encouraging all students to participate, (5) waiting patiently for students to collect their thoughts before speaking; a minute of silence can seem like a very long time, but research shows student responses are more complex and complete when there is adequate wait time.

Brain-Compatible Teaching

Why: Since 1987, research has greatly expanded our understanding of how the brain functions. The activities in this guide are brain-compatible (a term coined by Leslie Hart in *Human Brain and Human Learning*) because they incorporate strategies which make learning easier.

How: The brain-compatible activities in this guide:
- *Encourage collaboration between students.* The exchange of ideas fosters divergent thinking and creativity.
- *Engage students' feelings along with their intellect.* Research shows that learning and memory are enhanced when students have an emotional connection with the subject matter—the head and heart working together.
- *Connect the information to real life experiences.* Learning becomes meaningful when students make connections and use the information immediately in their own lives.
- *Co-create a rich learning environment.* When students contribute to creating visual displays, they are given the messages, "Your work is valuable," "What you do matters," or "This is important to others." Increased self-esteem results in greater risk-taking and participation, thus increasing learning.

A Drop Around the World Unit Plan

The three basic components of A Teacher's Guide to *A Drop Around the World* are 1) **Water Magic** lesson plans which focus on water and its importance, 2) **World Habitat Maps and Background Information,** and 3) **Skills for Living** lesson plans which focus on the development and practice of positive character qualities. Each component can be used individually or integrated into a six-week unit. The lessons work together, providing a variety of projects and activities to meet different learning styles.

Overall Unit Goals

- Encourage greater awareness and a feeling of connection with the physical world, especially the water component.
- Foster a vision of the orderliness of the universe and an appreciation for the ecological balance of all life.
- Inspire sensitive and harmonious relationships with the environment.

Benchmarks

In addition to the benchmarks indicated in individual lesson plans, the following Geography Benchmarks are met in this unit:
- Knows major physical and human features of places as they are represented on maps
- Knows the characteristics of a variety of regions
- Knows the consequences of specific physical processes operating on Earth's surface
- Understands the characteristics of ecosystems on Earth's surface
- Understands how human actions modify the physical environment
- Understands global development and environmental issues

Unit Objectives

Students will:
- Understand the water cycle and the importance of water to the Earth
- Research seven world habitats
- Make oral presentations
- Construct 3-D displays
- Create collages
- Design meaningful questions
- Practice Skills for Living

Elements of the Unit

1. Water Magic
Key concepts of the water cycle, water purification, and the importance of water to the human body and to our planet are concepts that are presented through demonstrations, activities, and projects. The topics of the lesson plans correspond to the topics presented at the end of *A Drop Around the World*. After reading the book aloud to the class, use the demonstrations from "Transformation Tricks" to introduce the unit.

2. World Habitats
A habitat consists of the plants, animals, and environment of a particular area. The habitats presented are some of the ones that "Drop" visits in its trip around the world. For each habitat there is a page of background information for the teacher which includes the basic characteristics of the habitat, a few common plants and animals that live in the habitat, and some of the ways people use and misuse the habitat. A Map Page is provided which can be given to students to color and label. A World Cities Map is also included which can be used to show which habitats support large concentrations of people.

An interesting way to introduce the habitats to students is through a habitat song (see page 11). Play the song while students follow along with the lyrics. After listening, ask students what habitats are found both in the song and in *A Drop Around the World*. (Ocean, desert, rainforest, grassland, mountain, Arctic and Antarctica.) Tell students that they will find out about more about each of the habitats in the song except grasslands. In addition, they will learn about coral reefs and coasts (which are found in *A Drop Around the World*).

Divide the class into seven small groups, one for each habitat. Provide a variety of reference materials and tell students that they should each research particular information about the habitat. This information includes habitat location, characteristics, plants and animals, and the way people interact with the habitat. Students can write down their information by making lists or by creating Mind Maps. Explain that they will use the information in four different ways:

A) Collage

Provide magazines (*National Geographic*, *Ranger Rick*, and *National Wildlife* magazines work very well) for students to cut up for pictures of their habitat. After assembling a wide variety, have students create a collage on one or more pieces of poster board.

B) Three-dimensional Display

Provide each group with a table or a particular area of the classroom where they can construct a display depicting the various aspects of their habitat. Have a wide variety of art materials including several kinds of paper, clay, papier mache, paints, and markers. Encourage students to use their own ingenuity and creativity to design a realistic representation of their habitat. They can bring in objects from home such as stuffed animals, shells, and plants. Making a diorama in a large box is an alternative to a display.

C) Oral Presentation

After gathering information, each student in the group will choose topics to present orally to the entire class. In addition to their collages and displays, they can use visual aids such as photographs, charts, artifacts, and foods of their habitat. Instruct students to write ten questions that can be answered by listening to their presentation. If the questions are given to the class before the oral presentation begins, the class will be able to focus their attention while listening. The following sample questions were written by fourth and fifth grade students: In what way do plants grow differently in the Arctic than in any other region? Compare the Arctic and Antarctic regions. Describe the relationship between a clownfish and sea anemone. What are the procedures a sea turtle goes through to lay her eggs? What characteristics do all starfish share?

By working in small groups students have many opportunities to practice interpersonal skills such as cooperation, communication, negotiation, compromise, decision making, encouragement, support, and appreciation.

Habitat Maps

The habitat maps can be used in a variety of ways:

- They can be used by the student group researching the habitat and included in their 3-D display.
- They can be presented throughout the unit to be colored and labeled by all students as a class assignment.
- They can be incorporated into the Water Magic Lesson Plans.
- They can be set up at a learning station, along with atlases and other reference materials, to be colored and labeled by students working independently and turned in at the end of the unit.

3. Skills for Living Lesson Plans

Water beautifully exemplifies many human qualities. For example, water in a river is vital, constantly moving towards the ocean; in the desert, water is appreciated because it is scarce. The many qualities of water relate easily to the Skills for Living. By connecting a Skill for Living with a quality students can see in action in the physical world, students may be better able to understand that quality in their own lives. The Skills for Living lesson plans are written to be used alone or as part of a unit.

Assessment

Although evaluating student progress and success can take many forms, the following criteria may be used for each of the different areas to be evaluated.

Habitat Research

- thorough, accurate, adequate information
- all information listed or Mind Mapped
- student stayed on task during research process

Habitat Maps

- labeled accurately
- colored neatly

Habitat Collage
- eye appealing
- thoroughly represents the habitat

3-D Display
- illustrates characteristics of the habitat
- creatively presented

Oral Presentation
- presented with poise and confidence
- thorough and accurate information
- well explained
- ten questions written and answered

Written Test
- compiled from group presentations and administered to the entire class

Suggested Six-Week Format

Week 1
Monday—Read *A Drop Around the World* and Water Magic: Transformation Tricks
Tuesday—Water Magic: Pollution Solution
Wednesday—Water Magic: Body Buddy
Thursday—Water Magic: Cloud Maker
Friday—Skills for Living: Tropical Rainforest

Week 2
Monday—Water Magic: The Green Machine
Tuesday—Introduce World Wide Habitats; student groups begin research
Wednesday—Aspiration: Mountains
Thursday—Continue research
Friday—Water Magic: Super Soaker

Week 3
Monday—Flexibility: Coasts
Tuesday—Continue research
Wednesday—Begin assembling materials for displays
Thursday—Continue research and work on the displays
Friday—Calmness: Polar Regions

Week 4
Monday—Awareness: Coral Reefs
Tuesday—Continue with research and work on the displays
Wednesday—Begin collages
Thursday—Continue working on research, displays, and collages
Friday—Vitality: Rivers

Week 5
Monday—Responsibility: Cities
Tuesday—Complete displays, finish collages, write questions
Wednesday through Friday—Water Magic: Temperature Tamer

Week 6
Monday—Practice oral presentations
Tuesday—Present oral presentations, displays, collages, brochures, questions
Wednesday—Open House for parents or other classes
Thursday—Written test
Friday—Appreciation: Deserts

Have to Have a Habitat

by Bill Oliver, Glen Waldeck and The Otter Space Band
(see Resources, page 48)

"Chorus"
Habitat, habitat, have to have a habitat.
Habitat, habitat, have to have a habitat.
Habitat, habitat, have to have a habitat.
You have to have a habitat to carry on.

The ocean is a habitat,
A very special habitat.
It's where the deepest water's at.
It's where the biggest mammal's at.
It's where our future food is at.
Keeps the atmosphere intact.
The ocean is a habitat we depend on!

"Chorus"

The desert is a habitat,
A very special habitat.
It's where the burning sands are at.
It's where the scarcest food is at.
It's where the brown kit fox is at.
Oasis keeps coolness intact.
The desert is a habitat we depend on!

"Chorus"

The rainforest is a habitat,
A very special habitat.
It's where the birds go calling at
And where the deepest roots are at.
Trees grow high in the canopy;
Make oxygen for us to breathe.
The rainforest is a habitat we depend on!

"Chorus"

A grassland is a habitat,
A very special habitat.
It's where the longest neck is found.
It's where the biggest nose sniffs around.
Where the bison get big and fat.
It's the home of the fastest cat.
Grasslands are a habitat we depend on!

"Chorus"

A mountain is a habitat,
A very cool habitat.
It's where the highest place is at.
It's where some springs and streams are at.
A mountain lion is a cat
that lives in this high habitat.
A mountain is a habitat that we depend on!

"Chorus"

Antarctica is a habitat,
A very special habitat.
It's where the highest glacier's at.
It's where the emperor penguin's at.
It's where the penguins must have fat.
It's where the coldest place is at.
Antarctica is a habitat that we depend on!

"Chorus"

The Arctic is a habitat,
A very special habitat.
It's where the Arctic wolf is at.
It's where the lands are cold and flat.
Glaciers keep the temperatures intact.
It's where the northern lights are at.
The Arctic should get a great big pat
For being a great big habitat.

"Chorus"

11

Water Magic  *Transformation Tricks*

Objectives	• Understand the water cycle (evaporation, condensation, precipitation)
Benchmarks	• Knows that the cycling of water in and out of the atmosphere plays an important role in determining climatic patterns; water evaporates from the surface of the Earth, rises and cools, condenses into rain or snow and falls to the surface, where it forms rivers and lakes and collects in porous layers of rock (Science 1, Level III)
Skills for Living	• Concentration
Materials	• Magician costume (cape, wand, white gloves); blackboard; chalk; paintbrush; fan; tea kettle; hot plate; glass jar kept in the freezer for an hour; jar holder (used in canning food); metal pie plate; hot pads; ice cubes; Water Cycle (Copy Master, page 47)—1 per student

Awaken Enthusiasm

Enter class dressed as a magician (a guaranteed way to engage the interest of the students).

Focus Attention

Tell the class that water can do many of the tricks that a magician can do.
- A magician can make things disappear; water can evaporate and disappear.
- A magician can make objects appear out of thin air; precipitation makes rain, snow, sleet, and hail appear out of thin air.
- A magician can change one object into another object, such as turning a scarf into a rabbit; condensation can change water from a gas into a liquid, such as changing water vapor into water droplets. Water can also change from a liquid to solid (when it freezes and becomes ice) and can change from a solid to liquid (when it melts).

Tell the class that they will have an opportunity to watch the magic of water through observing some simple demonstrations.

Direct Experience

Demonstration of Evaporation: With chalk, draw two circles at opposite ends of a blackboard. Use a clean paint brush to brush a thin coating of water on the inside of each circle. Set up a fan to blow onto one of the circles. Have the class observe what happens to the water. Ask them what action speeds up the evaporation process. (Moving air.) Another simple demonstration is to put two inches of water into a saucepan. Boil the water for several minutes. Measure the amount of water in the pan. Ask the class where the water went. Did they see it? (Water evaporating may be seen as steam.) Tell them that water is constantly being evaporated from the oceans, the rivers, lakes, ponds, and all other sources of water on Earth. There is ten times the amount of water in the Earth's atmosphere than in all of the rivers of the world combined.

Demonstration of Condensation: Explain to the class that the water vapor in the air changes to a liquid form when it is cooled, and forms very tiny water droplets which we see as clouds. Fill a tea kettle with water and heat it on a hot plate. Using a jar holder, hold a cold glass jar upside down above the spout of the tea kettle. Have students observe the water vapor rising and condensing as water droplets form on the sides of the cold glass. They can also observe condensation when they take a hot shower and the warm water vapor condenses on the cool bathroom mirror.

Demonstration of Precipitation: When the tiny water droplets of a cloud collide they stick together and form larger drops. When the drops become big enough and heavy enough they fall to earth as precipitation (rain, snow, sleet, or hail). Boil water in a tea kettle until steam (water vapor) comes out of the spout. Place several ice cubes in a metal pie plate and, using hot pads, hold the plate over the spout. Water drops will condense on the pie plate and, when the plate is tilted, they will fall as precipitation. Tell the students that one prediction is that without precipitation the Earth's oceans would dry up completely (evaporate) in 4,000 years. Explain that there is always the same amount of water on the Earth; and it is constantly changing form, going from a liquid to a gas back to a liquid and sometimes changing from a liquid to a solid and then back to a liquid. The process of water circulating from the oceans to the air to the land and back to the oceans is called the water cycle.

Share Inspiration

Look through *A Drop Around the World* page by page, identifying the parts of the water cycle represented on each page. Pass out copies of the Water Cycle handout (page 47), and have students color them.

Extension: Have students act out the water cycle. Divide the class into the various stages of the Water Cycle: 1) the oceans, rivers, lakes, 2) the process of evaporation including the sun and the wind, 3) condensation, the forming of clouds of all shapes and sizes which move around the world on wind currents, 4) various forms of precipitation including rain, snow, sleet, and hail. Give the groups time to get together and discuss how they want to portray their parts. Begin the performance of the "Water Magic Show" by having the first group come up to the front of the class and present their part of the water cycle. Add the other three one at a time until the entire water cycle is represented. (If the class is large, divide the class in half before designating the four groups and have the two groups present the water cycle to each other.) When finished, have students share appreciative comments with their classmates.

Water Magic 💧

Objectives	• Demonstrate two ways to purify water (filtration and distillation)
Benchmarks	• Knows that people have always had problems and invented tools and techniques to solve problems (Science 18, Level II)
Skills for Living	• Problem solving
Materials	• Two glass pitchers—one filled with muddy water, one filled with water that has red food coloring added; *for distillation demonstration—* a small tea kettle; small piece of modeling clay to seal off the spout of the tea kettle; 2 feet of rubber tubing; pan filled with ice cubes; 1 pint glass jar; a hot plate; *for filtration demonstration—*large funnel; glass jar with a mouth large enough to support the funnel; coffee filter; small pebbles; coarse gravel; fine sand

Awaken Enthusiasm

Display all of the equipment on a table making sure students can see and identify all of the items. Divide the students into groups of three to five students each.

Focus Attention

Show the students the two pitchers of water. Ask students to describe the procedure they would use to purify one of the pitchers of water using any of the supplies they see on the table. Allow 15 to 20 minutes for them to write up their demonstrations.

Direct Experience

Carry out two demonstrations for the entire class: 1) Purify the red water using a still. To make the still, attach the plastic tubing to the spout of the tea kettle, sealing off the spout with modeling clay. Place the pint jar into the pan of ice cubes and put the end of the tubing into the jar. Fill the kettle with the red water and place it on the hot plate. Turn up the heat to "high." While waiting to see what happens, proceed to the next demonstration. 2) Set the funnel into the mouth of the glass jar. Line the funnel with the coffee filter and layer in the following order: an inch of small pebbles, three inches of coarse gravel, and three inches of sand. Pour the muddy water into the funnel. Observe what happens. The result of the first demonstration will be that clear water will condense on the plastic tubing and drip into the jar, resulting in distilled water. Explain to students that distillation removes impurities, and that distilled water is used in science research, industry, and the home; even car batteries require distilled water. The result of the second demonstration will be that clear water will flow into the jar, resulting in filtered water. Water is naturally filtered by the earth as it percolates through soil, sand, and gravel layers. Point out to students that although filtration makes water cleaner, the water may still contain bacteria and be unsafe to drink. Read aloud the paragraph at the back of *A Drop Around the World* entitled "Pollution Solution."

Share Inspiration

Ask each group to identify the ways the demonstrations they planned were similar and different from the demonstrations that were actually conducted.

Extension: With additional time and materials students can conduct the demonstrations they planned and see how their results differed, or they can replicate the above demonstrations.

Water Magic

Objectives	• Know how much water is needed by the body
Benchmarks	• Knows healthy nutrition practices (Health 6, Level II)
Skills for Living	• Self-reliance
Materials	• An apple, a carrot, a head of lettuce, a picture of a human body; small slips of paper with 84%, 88%, 95%, 65%-70% written on them (an equal number of slips for each percentage)—1 slip per student; bathroom scale; water containers (either 8 oz. cups, 1 pint or 1 quart water bottles)—1 per student; several permanent markers

Awaken Enthusiasm

Display the apple, the carrot, the head of lettuce, and the picture of the human body in the front of the room. As class begins, have each student draw a slip of paper with a percentage written on it. Tell the class that one of the objects in the front of the room is composed of the percentage of water that is indicated on their slip. Have them guess which object their percentage corresponds with and, at a given signal, have students place their slip of paper in the front of the object. After all students have made their choice, read the percentages and indicate which percentage predominated for each object. Tell students the actual percentages are: an apple is 84% water, a carrot is 88% water, lettuce is 95% water, and the human body is about 65% to 70% water.

Focus Attention

Ask students how water is used in their bodies and make a list on the board. Emphasize how important it is to drink water and eat foods, such as fruits and vegetables, that have a high water content.

Direct Experience

Have students figure out mathematically how much water their body needs per day, as recommended by the EduK Foundation. Students first need to know how much they weigh in pounds. They can weigh themselves or work with a partner. Because the human body requires one ounce of water for every three pounds of body weight, they should divide their weight by 3 to get the number of ounces of water they need each day. That number, the quotient, is then divided by 8 to get the number of 8 ounce glasses of water they need to drink each day. Students can follow the formula:

Body weight divided by 3 = number of ounces of water per day

Number of ounces of water divided by 8 = number of glasses of water per day

A simpler, although a little less accurate, way to figure the amount of water needed is to figure one pint of water (two 8 ounce glasses) for every 50 pounds of body weight.

Share Inspiration

Supply students with a water container—either an 8 ounce cup (appropriate for younger students) or 1 pint or 1 quart water bottles. Have students use permanent marker to write their names and decorate their water container. Have everyone fill their container with water and take a big drink. Remind students to drink throughout the day. Drinking water regularly is an important habit for all people to develop.

Water Magic

Cloud Maker

Objectives	• Use imagination and creativity to write a cloud-related vertical poem
Benchmarks	• Demonstrates competence in expressive writing (Language Arts 1, Levels II and III)
Skills for Living	• Creativity, cooperation
Materials	• Plain white paper, blue construction paper, notebook paper—1 sheet of each per student; tape—2 pieces per student; classical music tape

Awaken Enthusiasm

Notice the clouds outside of the classroom window. As a class or in small groups, brainstorm a list of adjectives that describe the size, shape, texture, and color of clouds they see; for example, wispy, fluffy, soft, dark. Expand the list to describe other clouds they have seen.

Focus Attention

Play the music tape softly and instruct students to close their eyes and imagine that they are looking up at the sky. Tell them that clouds are forming into many different shapes; some might look like animals or people or unusual objects. Have students use their imaginations for a minute or two as they watch the clouds float by, changing shapes. Turn off the music and have them open their eyes. Using the plain white paper, have them tear it into one of the shapes they "saw." (Reassure them that it's O.K. if they didn't see any shapes or if they have another idea of a shape they want to use; the objective is simply to have them tear out a cloud shape.) Once their shape is torn out, have them glue it to the blue construction paper. Next have them tape the piece of notebook paper on the reverse side of the construction paper at the top of the page leaving the bottom of the page free. When finished, instruct each student to pass his paper to another student who looks at the shape of the cloud. Then that student turns the paper over and writes down what he thinks the shape looks like on the bottom line of the notebook paper. He hides his writing by folding up the bottom edge of the paper. He then passes the picture to someone else who follows the same procedure of looking, writing, and folding.

Direct Experience

After eight passes, tell students to return the paper to the original student who reads what others have "seen" in his cloud shape. Tell students to choose either their own idea, or one from the notebook paper, to be the subject of a vertical poem. To write a vertical poem, instruct students to write the name of the object vertically down the left margin of a piece of paper. Each letter of the word becomes the first letter of a word which begins a line in the poem. For example, the word "lion" becomes:

    L arge and ferocious
    I t fills the sky
    O nly to be blown into a
    N ew shape by the wind.

Share Inspiration

Check students' spelling and have them write their poems neatly onto the construction paper. Display clouds and poems around the room.

Water Magic

Objectives	• Identify unique water features of the Earth as the "water planet"
Benchmarks	• Knows that life is adapted to conditions on Earth, including the strength of gravity to hold an adequate atmosphere and an intensity of radiation from the Sun that allows water to cycle between liquid and vapor; knows that nine planets of differing sizes and surface features and with differing compositions move around the sun (Science 3, Levels III and IV)
Skills for Living	• Creativity
Materials	• Plain white paper (preferably 11" by 14"); colored pens or pencils; magazines that can be cut up; travel brochures; nine index cards; pictures of the planets of our solar system; tape • Teacher Preparation: Write three to five facts about each planet using one index card for each planet. • Time: Allow 2-3 class sessions. (This activity can be used as a culminating activity for a six-week unit. Student groups would present water information about their chosen habitat.)

Awaken Enthusiasm

Ask students to identify the planets of our solar system. List them on the board. Show pictures of each of the planets. Divide the students into nine groups. Give each group one of the index cards containing facts about one planet. Instruct them to read the facts and decide to which planet they correspond. Have students bring their card to the front of the class and tape it next to the correct planet. Correct any errors and read aloud all of the facts.

Focus Attention

Ask students what similarities they notice between Earth and the other planets. What differences do they notice? Point out that a big difference is that Earth contains water. Read and/or Mind Map the information entitled "Temperature Tamer" at the back of *A Drop Around the World*.

Direct Experience

Instruct student groups to design and make a travel brochure advertising Earth to the inhabitants of the Milky Way Galaxy. Have each group emphasize places of interest on Earth that include water. Show students sample travel brochures to give them examples about how places of interest are featured. Have them use *A Drop Around the World*, atlases, geography books, and magazines to find unique and unusual places and ideas. Remind them that they can use "frozen water" to focus on such places as the Alps, Antarctica, or Alaska. Original art work as well as pictures cut from magazines can be used to create the brochures.

Share Inspiration

When finished, have groups present their brochures to the class as if they were travel agents. After the presentations, lead a class discussion identifying three or more creative or well-done features about each brochure.

Water Magic

Objectives	• Play an active water-related relay game
Benchmarks	• Understands basic Earth processes (Science 2, Level II); works in a group to accomplish a set goal in both cooperative and competitive activities (Physical Education 5, Level II)
Skills for Living	• Vitality
Materials	• Buckets of equal size—2 per group; large sponges—1 per group; a source of water to fill buckets; an outdoor playing space
	• Teacher Preparation: Set up the playing area by placing empty buckets (one for each five- or six-person team) five feet apart in a straight line. Place a sponge next to each bucket. About ten yards away from each empty bucket, set up a bucket filled halfway with water.

Awaken Enthusiasm

Divide the class equally into small groups of five to six students. Ask each group to brainstorm as many answers as they can to complete the sentence, "The earth is like a sponge because......." After two minutes of brainstorming, have each group choose one answer to share with the rest of the class.

Focus Attention

Read aloud the paragraph from the back of *A Drop Around the World* entitled "Super Soaker." Refer to the page of the book showing the well to illustrate how the water soaks deep into the earth.

Direct Experience

Explain to the class that they are going to play a relay game outside. Once outside, instruct each team to line up behind an empty bucket and give them the following directions: At a given signal, the first person in line should pick up the sponge, run to the bucket of water across from their team, soak the sponge with water, run back to their team's empty bucket, wring the water out of the sponge, and hand the sponge to the next person in line. The relay continues until the original water-filled bucket is empty. There are two ways to win the relay: be the first team to empty their bucket, or be the team that transfers the most water from one bucket to the next.

Share Inspiration

Have team members share appreciations with others in their group, noticing such qualities as determination, energy, and speed. Use the water in the buckets to water plants around the school campus.

Water Magic

Objectives	• Grow sprouts as an example of a "green machine"
Benchmarks	• Knows that plants depend on light and water; knows that almost all food energy ultimately comes from the Sun as plants convert light into energy (Science 8, Levels I & III)
Skills for Living	• Practicality
Materials	• Quart-size glass jars—1 per student; fine mesh screen or cheesecloth—a piece to fit over the mouth of each jar; strong rubber bands—1 per jar; bowls large enough to hold the jars at an angle when they are inverted; sprouting seeds: 1 tablespoon alfalfa seeds, 1/3 cup garbanzo beans, 1/3 cup lentils, 1/3 cup mung beans per jar; a sink; a table or shelf to hold all of the jars; masking tape; small plastic bags and ties—1 per student; choice of salad dressings
	• Teacher Preparation: Measure the appropriate amounts of sprouting seeds or beans into small plastic bags and secure the tops with a twist tie. Place all of the small bags into a grocery bag.

Awaken Enthusiasm

Have students reach into the grocery bag and draw out a small plastic bag of seeds or beans. Have students form a group with other students who have the same kind of seeds or beans. Ask them to guess what kind of seeds or beans they have.

Focus Attention

Identify the types of seeds or beans each group has. As students return to their desks, instruct them to take a jar from the table and fill it with water. They should also pick up a piece of screen and a rubber band. Have them add their seeds to the water and use the rubber band to secure the screen over the top of the jar. Have them use masking tape and a pen to label their jar with their name. When finished, the jars should be placed on the designated table. The seeds need to soak overnight.

Direct Experience

The next day, have students drain the water from their jars and rinse their seeds by adding fresh water to the jar and then dumping it out. The jars of rinsed sprouts can rest inside of a bowl at an inverted angle to allow all water to drain away. Have students notice any changes they see in the seeds. Sprouts should be rinsed twice a day. Depending on the temperature of the room, the lentil, mung, and garbanzo sprouts will be ready after only three days and can be placed in the sun to green for a few hours before being put in a container and refrigerated. Allow an additional two days for the alfalfa seeds to be fully sprouted.

Share Inspiration

Combine all of the sprouts into a big salad bowl and have students use their sprouting bowls for individual servings. Having a choice of tasty salad dressings can encourage any reluctant eaters to give the sprout salad a try. While enjoying their salads point out to the class a few of the practical advantages of growing and eating sprouts: they can be grown year-round inside, are rich in nutrients including protein, can be eaten raw or cooked, and are an inexpensive, nutritious, easy-to-grow source of food.

Description: Coasts occur where the land meets the water. Two of the most important factors that shape coasts are tides and waves. Coasts are usually sandy beaches or rocky shorelines. Wetland areas (salt marshes and swamps) often stretch along coastlines.

Characteristics:
- Constantly changing conditions
- Battering waves
- Regular fluctuations of high and low tides
- Intertidal zones including tide pools

Plants:
- Adapted to constant change and movement
- Marine algae and seaweeds including kelp, sea palms, and rockweeds
- Beach grasses

Animals:
- Adapted to constant change and movement
- Animals with shells including limpets, mussels, crabs, clams, oysters, and barnacles
- Soft-bodied animals including jellyfish, anemones, and sponges
- Fish such as salmon and sea bass
- Shore birds such as sand pipers, cormorants, seagulls, and herons
- Sea mammals including otters, seals, sea lions

Human Connection:
- People dump garbage into the ocean, especially in the form of plastics.
- Oil slicks from boats and industry poison sea creatures and interfere with animal survival.
- Over-fishing upsets the ecological balance.
- Sewage dumped into the ocean contaminates water.
- Runoff from coastal cities and highways pollutes water.
- Managed aquaculture and marine culture lead to an increase in food supplies.
- Creation of laws, such as those limiting development, protects coastal environment .
- Programs have been developed for the protection of marine life such as turtles, otters, and seals.
- Sea water can be purified for human use.
- Wave action can be harnessed to produce energy.

World Habitat Map—Coasts   (Label and Color)

Pacific Ocean	North America	Asia
Atlantic Ocean	South America	Australia
Indian Ocean	Europe	Antarctica
Arctic Ocean	Africa	

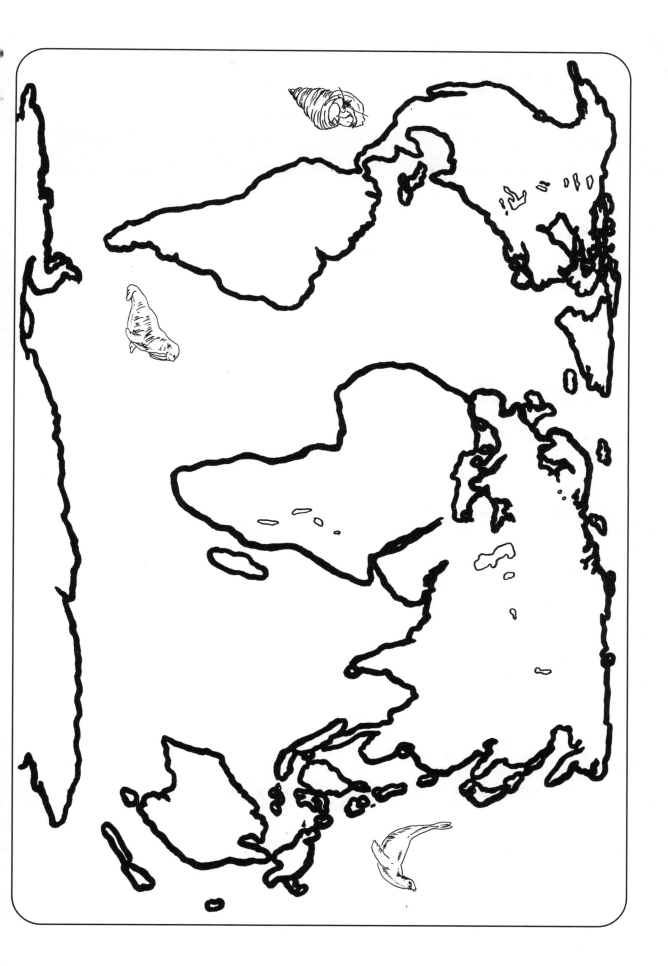

Description: Mountains are land masses that rise much higher than surrounding land. They are the result of strong upward and/or folding movements of earth and rock. Some are the result of volcanic activity.

Characteristics:
- Measured by height above sea level
- May occur in groups called ranges; ranges can be several hundred miles long
- Found on every continent
- Rise from deserts, plains, forests, and oceans
- Air gets thinner as altitude increases
- Temperatures drop 5.4 degrees Fahrenheit for every 1,000 feet in altitude in clear weather
- Appearance varies greatly depending on age—old mountains are rounded by wind, rain, and erosion; new mountains have sharp, rocky tops
- Highest mountain: Mount Everest (Asia)—29,028 feet
- Longest mountain range: Andes (South America)—4,474 miles

Plants:
- Plant variety and growth depends on elevation, temperature, sunlight, wind, and rainfall.
- The timber line is the altitude on mountains above which trees stop growing.

Animals:
- Many live on mountain plants; others are predators
- Vary with elevation and climate
- Includes mammals such as marmots, ground squirrels, goats, sheep, yaks, deer, cougars, bears, and wolves
- Birds such as eagles, condors, hawks, vultures
- Many varieties of snakes and insects

Human Connection:
- Minerals such as coal, silver, gold, copper, and tin are mined.
- Timber is harvested.
- Erosion occurs due to deforestation.
- Popular mountain recreation activities include hiking, camping, and skiing.

World Habitat Map—Mountains

(Label and color)

North America	Europe	Asia	Australia
Rocky Mountains	Alps	Caucasus	Great Dividing Range
Appalachian Mountains	Pyrenees	Ural Mountains	
South America	Africa	Himalayas	
Andes Mountains	Ethiopian Highlands		

Description: A river is any natural stream of fresh water that flows in a channel. Rivers originate from ponds, lakes, springs, melting snow, mountain runoff, or streams. The amount of precipitation in an area determines the amount of water in that area's rivers.

Characteristics:
- Drain the land (carry away extra rain water) in a river basin (watershed)
- Flow from higher to lower elevations
- Change the land—carve out gorges and canyons, create valleys and islands
- Erode rock and soil
- Sometimes overflow their banks (flood) when there is too much rain or melting snow
- Small rivers (tributaries) flow into larger rivers
- Waterfalls tumble over cliffs

Plants:
- Underwater plants
- Water thirsty plants such as cattails and reeds
- Algae, mosses, and lichens

Animals:
- Water insects including dragonflies, mayflies, and mosquitoes
- Crustaceans such as crabs
- Amphibians such as frogs and toads
- Mammals that live in the water (such as otters) and visit the water (such as deer)
- Water birds that swim (such as ducks), wade (such as herons), and catch fish (such as kingfishers)
- Fish including trout, bass, and catfish

Human Connection:
- People use rivers for transportation, irrigation, drinking water, and generating energy.
- They are dammed to prevent flooding.
- Pollution is often caused by toxic run-off and erosion due to human development.

World Habitat Map—Rivers (Label and color)

North America	South America	Asia
Mississippi River	Amazon River	Volga
	Europe	Ganges
	Danube River	Indus
	Thames	

Africa
Niger
Yangtze
Nile
Australia
Darling

Zaire (Congo)

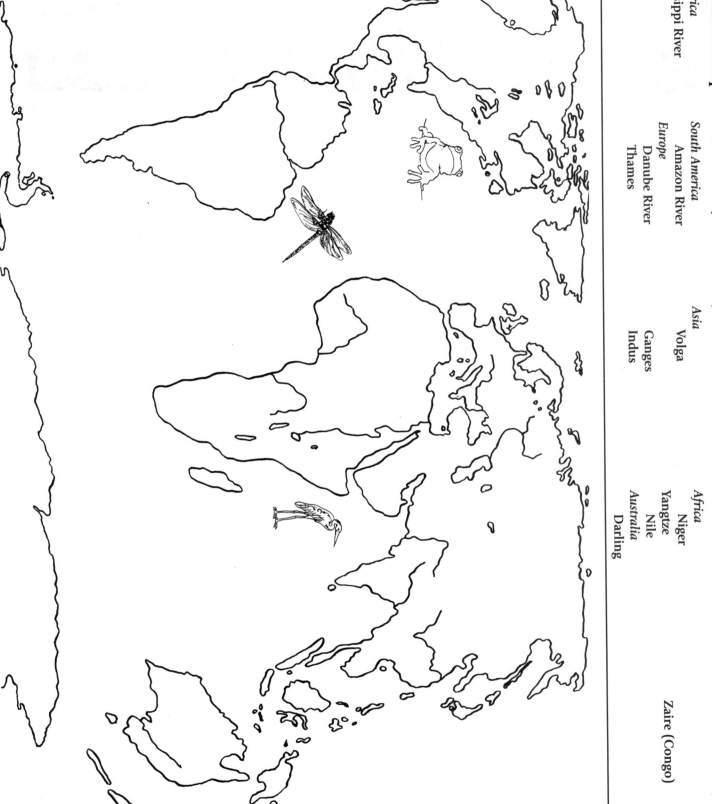

25

Description: A desert is a land where water is scarce because the climate is either too dry or too cold. There are dry deserts such as the Sahara, and cold deserts such as Antarctica. This section focuses on dry deserts; for information about Antarctica, see page 32.

Characteristics of Dry Deserts:
- Less than ten inches of rainfall a year
- Evaporation rate higher than the annual amount of rainfall
- Hot and dry air
- Little cloud cover
- Strong winds
- Sporadic rainfall that comes in bursts and then not at all
- Variable temperatures—hot days and cold nights
- Unusual variety of land forms such as rocks, cliffs, canyons, sand dunes, and salt flats
- Little vegetation
- Found on all continents except Europe
- Cover one quarter of the landmass of Earth
- Largest desert: Sahara (Africa)—3,243,240 square miles
- Driest desert: Atacama Desert (South America)—no rain for 400 years in one location
- Hottest Desert: Sahara (Africa)—136.4 degrees Fahrenheit recorded in Libya

Plants:
- Scrubby appearance; often dull green or gray in color
- Scattered grasses
- Some shrubs
- Cactus and succulents

Animals:
- Search for water predominates lives of many desert animals
- Rodents including mice and rats
- Reptiles such as snakes and lizards
- Mammals including camels, foxes, and kangaroos
- Birds such as owls, vultures, quail, and roadrunners
- Insects including scorpions and spiders

Human Connection:
- Native peoples often follow nomadic lifestyles.
- Irrigation is used for agriculture.
- Use of ground water supplies can easily cause the sources to dry up.
- In places, overgrazing causes erosion.
- Off road vehicles tear up plants and soil.
- Exploration and drilling of oil poses environmental threats to animals and their habitat.

World Habitat Map—Deserts

(Label and color)

North America	Asia	Africa	Australia
Mojave	Arabian	Sahara	one third of continent is desert
South America	Gobi	Namib	Antarctica
Atacama			entire continent

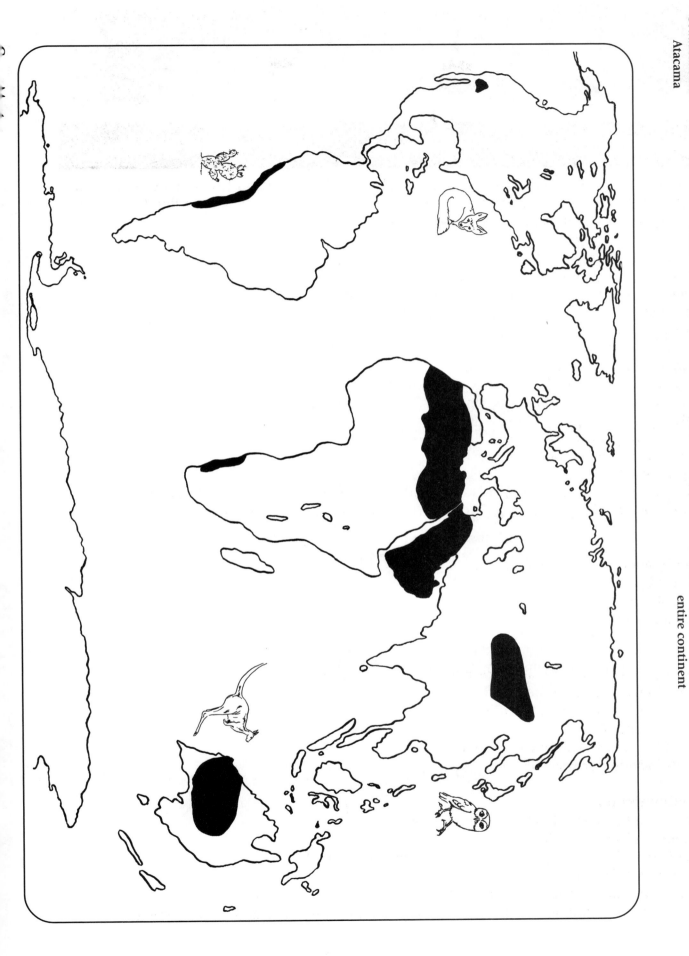

Description: Tropical rainforests are areas that receive a minimum of 80 inches of rainfall a year. They circle the world at 20 degrees of latitude on either side of the equator. (At higher latitudes, temperate rainforests exist such as the one on the Olympic Peninsula in Washington.)

Characteristics:
- Up to 200 inches of rain a year in some rainforests
- Temperatures between 70 and 85 degrees Fahrenheit
- Little difference between daytime highs and nighttime lows
- High humidity: 70% during the day, 95% at night
- Streams and rivers criss-cross the landscape
- Shallow soil poor in nutrients
- Very fragile
- Half of all of the plant and animal species of the world live in rainforests (biodiversity)
- Most rainy days: Mt. Wai'ale'ale (Hawaii)—350 days of rain per year

Plants:
- Adapted to large amounts of water
- Tall, broadleaf, and evergreen trees
- Trees often have buttress roots for support
- Many epiphytes, such as bromeliads, grow in trees
- Layered vegetation: emergent trees, canopy, and forest floor
- Many can grow in low light conditions

Animals:
- Lots of insects, especially ants
- Life is layered to reduce competition and increase survival
- Plants and animals form unique partnerships such as the sloth and algae
- Have developed striking examples of camouflage such as the jaguar

Human Connection:
- Deforestation is increasing yearly.
- Uncontrolled slash-and-burn agricultural practices destroy the rainforest.
- Inexpensive tropical beef is in demand in some countries which leads to deforestation in creation of pasture land.
- Future medicines may be derived from rainforest plants.
- Tourism is increasing.
- People using rainforest products can help sustain the rainforest.
- Many conservation groups have formed to protect the rainforest.

World Habitat Map—Tropical Rainforests

(Label and color)

Central America
South America
Africa
Southeast Asia
Australia

Description: Coral reefs, built by stony (hard) corals, are located in tropical ocean regions between the Tropic of Capricorn and the Tropic of Cancer. Corals are tiny animals that live in colonies and use the calcium and carbonate from sea water to make limestone which they use to build protective rock cups around themselves, creating a rock mass. New corals grow over old corals building the rock mass into a reef.

Characteristics:
- Takes thousands of years to build a coral reef
- Busy, crowded, colorful
- Provide living spaces for many other animals
- Protect coastal areas
- Water is above 68 degrees Fahrenheit
- Water is clear and shallow enough for light to penetrate
- Often form around small islands and eastern shores of continents
- Largest reef: Great Barrier Reef (Australia)—over 1000 miles long

Plants:
- Algae, a key to reef formation

Animals:
- Inhabited by a large number of species
- Animals share resources by maintaining different schedules
- Soft-bodied animals such as anemones and sponges
- Soft and hard corals
- Large variety of fish including clownfish, angelfish, groupers, and sharks
- Crustaceans such as urchins and crabs

Human Connection:
- Tourists use reefs for recreation such as snorkeling and diving.
- Reefs provide food such as fish, shrimp, and crabs.

World Habitat Map—Coral Reefs

Pacific Ocean	North America	(Label and color)
Atlantic Ocean	South America	Asia
Indian Ocean	Africa	Australia

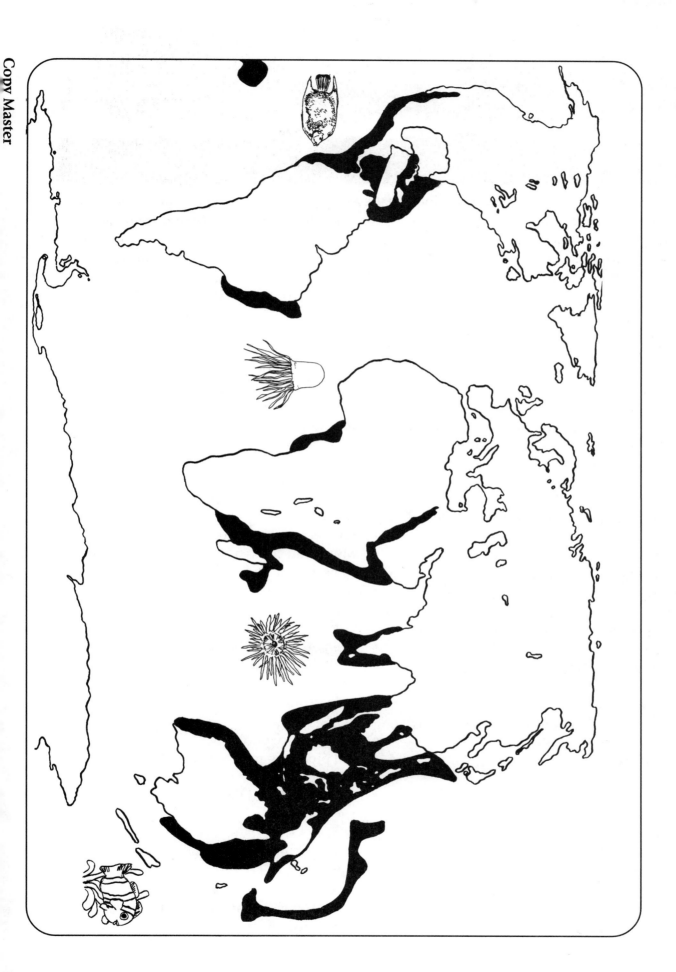

Description: The Polar Regions of the Earth include the Arctic and Antarctic. The Arctic is a region that consists of the Arctic Ocean, which surrounds the North Pole, and the northern parts of the continents of North America, Europe, and Asia. Its boundary is measured in different ways. Sometimes it is defined as the area north of the Arctic Circle. Other definitions include: the point where the tree line stops (where no trees will grow), and the place on land where permafrost (permanently frozen land) begins. Antarctica, which is bigger than the US. and Mexico combined, is the continent that surrounds the South Pole, including the ice shelves at the continent's edges (some as large as the state of Texas).

Characteristics:
Arctic
- Sun does not set for one or more days in June
- Sun does not rise for one or more days in December
- Blanketed with snow and ice for much of the year
- Most snow and ice melt in summer
- Land of extremes
- Freezing temperatures, average winter temperature is -30 degrees Celsius
- Severe winds
- Ice fields and ice bergs
- Snowy deserts

Antarctica
- Surrounded by three oceans—Pacific, Atlantic, Indian
- Covered with permanent snow and ice—up to 3 miles thick at the South Pole
- Winds up to 200 miles an hour
- Extreme light and dark period, similar to the Arctic but opposite seasons
- Coldest place on Earth

Plants:
Arctic
- More than 450 varieties
- Live frozen under the snow for 10 months
- Adapted to extreme conditions
- Lichen, algae, and mosses
- Tundra—covered in grasses, wildflowers, and low-growing plants

Antarctica
- Grow only along coast closest to South America
- Primitive plants including algae, lichens, and moss; only one type of grass

Animals:
Arctic
- Mammals including polar bears, seals, walrus, whales, foxes, hares, caribou, muskox, reindeer, wolves, and lemmings
- More than 100 types of birds including loons, terns, eagles, and owls
- Rich in fish such as salmon, cod, and rockfish

Antarctica
- Animals can live only along sea coasts
- Seals, whales, and penguins

Human Connection:
Arctic
- It is a source of oil.
- Oil pipelines and oil spills hurt the environment.
- The air and land become polluted by industry.

Antarctica
- No permanent human population exists on Antarctica.
- Over 40 countries operate research stations.
- Parts of the land are claimed by many different countries.

World Habitat Map—Polar Regions (Label and color)

The Arctic Region
North Pole
The continent of Antarctica
South Pole

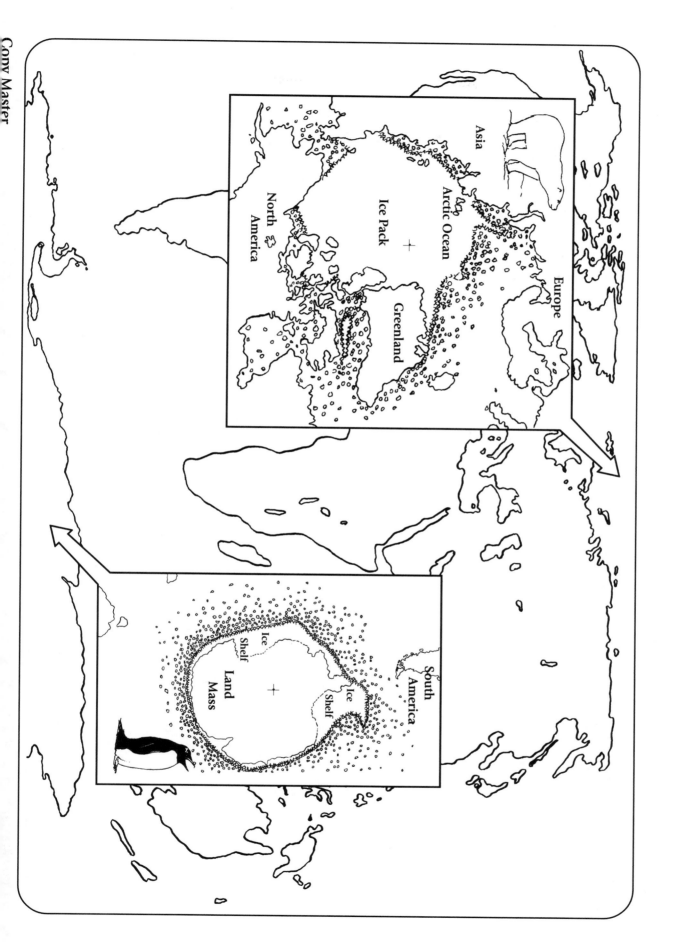

Asia

North
America

Arctic
Ocean

Ice Pack

Greenland

Europe

Ice
Shelf

Land
Mass

Ice
Shelf

South
America

33

World Habitat Map—Cities

(The cities identified on the map, except for Sydney, Australia, have populations of five million or more people.)

North America
New York, USA
Chicago, USA
Los Angeles, USA
Mexico City, Mexico

South America
Lima, Peru
Rio de Janeiro, Brazil
Buenos Aires, Argentina

Europe
London, Great Britain
Paris, France
Africa
Cairo, Egypt

Asia
Moscow, Russia
Beijing, China
Calcutta, India
Tokyo, Japan

Australia
Sydney

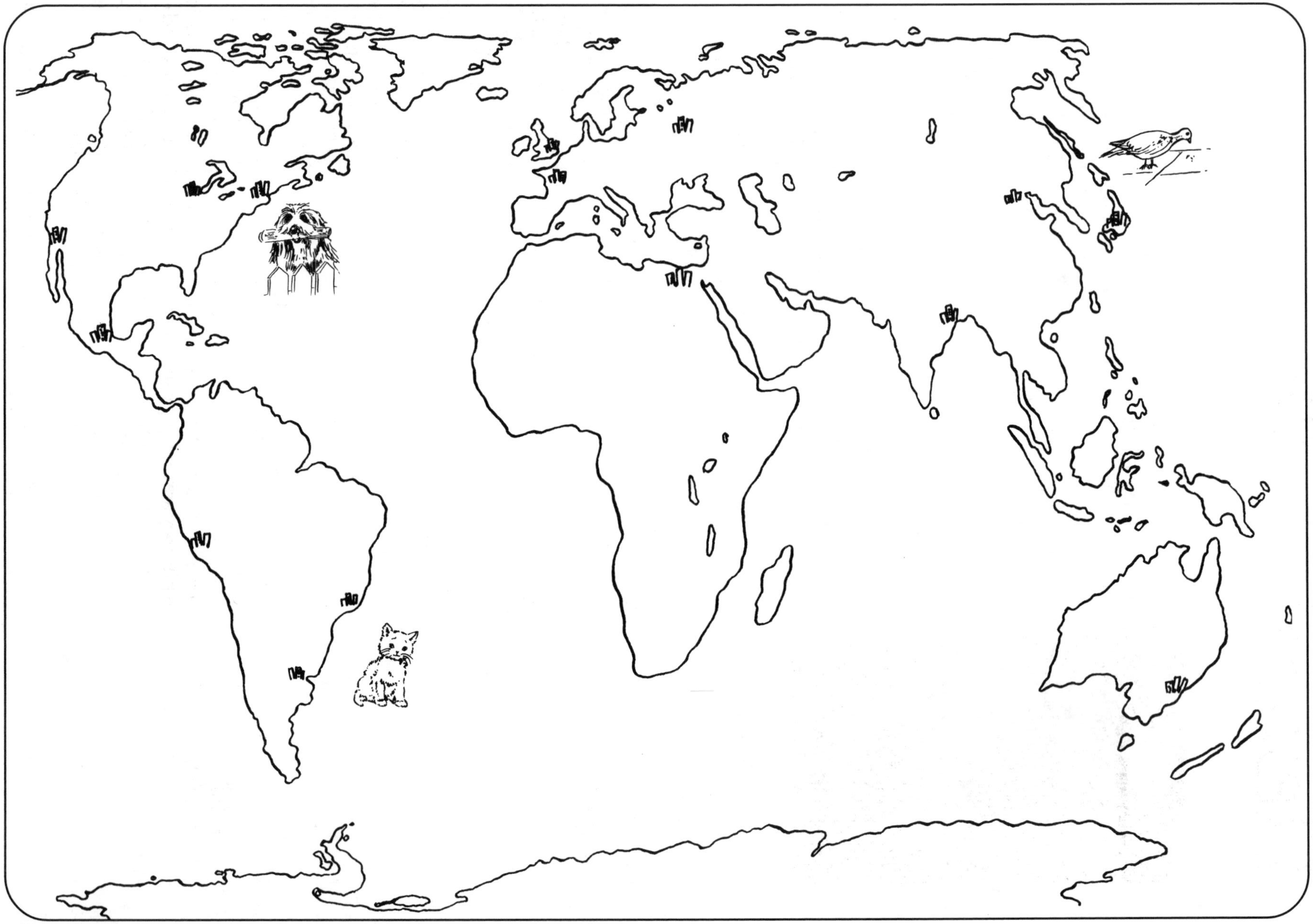

Flexibility

Water Is Changing

Objectives	• Identify personal changes • Experience being flexible
Benchmarks	• Understands a variety of physical, mental, emotional, and social changes that occur throughout life, and how these changes differ among individuals (Health 10, Level IV)
Skills for Living	• Flexibility
Materials	• Drawing paper—1 piece per student; crayons, markers, or colored pencils—enough for each student • Teacher Preparation: Make several big and small changes around the classroom such as moving a table, putting up a different poster, changing the bulletin board, or rearranging some items on the top of your desk.

Awaken Enthusiasm

Ask students to notice all of the changes in the room that have happened since the day before. (They may notice some you didn't intend.)

Focus Attention

Tell the class that one ecosystem where there is a lot of change is the coast, where the land and sea meet. Ask students to describe their experiences at the coast. What were the sights, smells, and sounds they noticed? What changes occurred at the coast while they were there? Tell the class that the water level of the ocean is constantly changing; there are times of low tide and times of high tide. The weather can also change quite quickly and dramatically. The plants and animals that live in a coastal environment have adapted to all of the changes. Some examples include the mussels that open and close, the limpets that cling tightly to the rocks, and the sea grass that easily bends in the wind. Explain that change is constant in most peoples' lives, too, no matter where they live. Ask students what changes have happened to them over the last year. For example, did they get a new pet, join a sports team, or change a grade at school? Did their bodies change? Have they expeperienced any family changes—a death, divorce, new step parent, new baby? Learning to be adaptable and flexible is a healthy way to handle changes that come to them.

Direct Experience

Tell the class that they are going to have an opportunity to practice being flexible. Give each student a piece of drawing paper and coloring supplies. Tell them that they will each draw a picture of a coastal scene. (You may want to have several pictures of the seashore and coast available for them to look at.) Tell them that while they are drawing you will ring a bell at different intervals. Each time they hear the sound of the bell they should change pictures with another student and continue drawing on that picture. They will be practicing flexibility because they don't know when you will ring the bell and they will not be finishing their own picture. In fact, they will be working on many different pictures.

Share Inspiration

When finished, display the pictures around the room. Congratulate students on their ability to change and to be flexible.

Aspiration

Water Is High

Objectives	• Set and accomplish an academic goal
Benchmarks	• Sets and manages goals (Life Skills: Self-Regulation 1, Level IV)
Skills for Living	• Aspiration
Materials	• Aspiration Chart (Copy Master, page 37)—1 per student; optional: construction paper flag—1 per student; colored pencil—1 per student
	• Time: Daily for one or more weeks. (This activity may be used to set and manage the various assignments in a six-week unit.)

Awaken Enthusiasm

Ask the class: What skills does it take to climb a mountain? (Include physical, mental, and emotional skills.) How long does it take to train to get ready to climb a mountain. (Depending on the initial fitness of the climber, it can take one or more years of preparation for some expeditions.) Relate a few inspiring mountain climbing stories, such as the story of the group of handicapped climbers who climbed Mt. McKinley, or the man who trained for years to get to the top of the Matterhorn. *Reader's Digest* is a good source of real-life adventure stories. Fictionalized stories are also inspiring, such as the classic *Banner in the Sky.*

Focus Attention

Explain to the class that every person who climbs a mountain has a specific goal: getting to the top. Every plan they make and every step they take is with that goal in mind. Ask the students to brainstorm academic (school work) goals that they would like to accomplish in the next week (or the next several weeks). Goals are most effective when they are specific, concrete, and measurable. For example, a goal to "earn a 90% on a spelling test" is much more specific than a goal to "learn how to spell." Other examples of academic goals include finishing reading a book, memorizing the multiplication facts for the number 8, or writing the final draft of a report. Give each student a copy of the handout and instruct him to write his or her goal at the top of the Aspiration Chart.

Direct Experience

Each day have students write down the action they took towards accomplishing their goal on one of the ladder segments on the Aspiration Chart. (For example, practiced spelling words with a friend, used flash cards for 5 minutes, wrote two paragraphs.)They can then lightly color in that part of the ladder with a colored a pencil. (If using this activity in conjunction with the six-week unit, the goal for all students can be to complete their project. Written on the ladder can be the various steps of researching their habitat such as researching information, preparing their oral report, writing their final questions, making their 3-D display, and presenting their project to the class.)

Share Inspiration

Display the students' Aspiration Charts around the room as a reminder of their accomplishments. Optional: Present each student with a flag when they reach the top of the mountain (meet their goal). The flag can be made out of a piece of colored construction paper attached to a straw or strip of cardboard. Write the words "Congratulations! You accomplished your goal!" on the construction paper.

Goal

Copy Master

Vitality

Objectives	• Interpret positive words and phrases with appropriate movements
Benchmarks	• Understands physical activity as a vehicle for self-expression (Physical Education 5, Level III); creates dance that communicates topics/ideas of personal significance (The Arts: Dance 3, Level III)
Skills for Living	• Vitality
Materials	• Tape of instrumental music that is quietly energetic such as Vivaldi or Mozart

Awaken Enthusiasm

Instruct the class to make a large circle around the room. Play the music and have the class walk around to the tempo of the music, staying in a circle. While continuing to walk in a circle, tell the students to move their bodies like a tiny stream that is barely trickling down from a hillside of melting snow. Tell them to express through their body movements the stream becoming a flowing river. Have the river gradually slow down as it gets colder until it finally freezes (stops), but then it begins to melt and starts to flow once again. The river continues to freeze and thaw, with the movement decreasing and increasing. End by saying these words from *A Drop Around the World*, "Free and fluid flowing fast; free and liquid, loose at last!" as the students move quickly; tell them to move towards their individual desks. Say, "It stops at a spigot, meets a dead-end." Tell students to sit down at their desks.

Focus Attention

Ask students what movements they either did themselves or noticed others doing that *really* expressed the different aspects of a river. Tell them that body movements are a very powerful way to express meaning. When movement is combined with words, it can increase the meaning of the words. Write the following three phrases on the blackboard: *I am awake and ready! I am tired and lazy. I reach for the stars.* Tell the class you are going to express one of the phrases with body movements and ask them to guess which phrase you are using. Vigorously march in place, pumping your arms forward and backward. Take the students' guesses. Have them join you saying aloud, "I am awake and ready!" several times while you march in place together.

Direct Experience

Divide the class into small groups of three to five students. Have each group choose a positive phrase to express through movements that they can teach to the rest of the class. The groups can make up their own phrases or they can choose from a list which you provide. For example, *I reach for the stars. If I think I can, I can! I say YES to life! Nothing can stop my progress! I feel so glad; I feel so good! I am brave, I am strong! I am a friend to everyone!* Give groups five to ten minutes to create their movements.

Share Inspiration

Have each group come up to the front of the class to teach their positive phrase and movements to the rest of the class. Then have the entire class do the movements together while repeating the words. When finished, applaud each group for its vitality, creativity, and self-expression. Use the words and movements throughout the year whenever the class energy needs to be changed.

Appreciation

Water Is Appreciated

Objectives	• Recognize and share appreciations with others in the class
Benchmarks	• Recognizes the contributions of others (Life Skills: Working with Others 5, Level IV); knows behaviors that communicate care, consideration, and respect of self and others (Health 4, Level II)
Skills for Living	• Appreciation
Materials	• Constructions paper—1 sheet per student; small slips of paper—1 per student; a small box to hold the paper slips; optional: a Certificate of Appreciation for each student • Teacher Preparation (optional): Make Certificates of Appreciation • Time: Two class periods or a 90 minute block

Awaken Enthusiasm

Show the class the picture of the desert from *A Drop Around the World.* Have students imagine a desert scene by closing their eyes and listening to the following description: "You are walking through the hot, dry desert. There is nothing but sand as far as you can see. The sun is shining brightly without a cloud in the sky. The burning sand is blown into your eyes by the wind whipping across the sand dunes. Your lips are parched and cracked and you are thirstier than you can ever remember. You begin to hear a faint sound that reminds you of a tinkling wind chime. Then you come to the top of a sand dune and see below lush, green grass surrounding a small, bubbling spring of water. You realize it is an oasis, and you run down to it. You eagerly bend down and take a long drink, splashing water on yourself."

Focus Attention

Ask students how it felt to drink and splash in the water. Tell the class that in the desert water is really appreciated. Then tell students that in the classroom they are really appreciated. One at a time, have students come up to the front of the class while you share an appreciation for each one. (You may want to give them a Certificate of Appreciation at this time.) Appreciations can be based on the Skills for Living, page 41. Mention a specific time when the student demonstrated that Skill for Living. For example, "I appreciate Jane for her even-mindedness. She didn't get upset even when water was spilled all over her finished art project. She just cleaned up the mess and did her best to fix her project without complaining or blaming anyone." (By giving specific examples, you can reinforce the positive behaviors you would like to see in class.)

Direct Experience

Explain to the class that you are not the only one who appreciates students in the class; they can appreciate each other. Pass out a small slip of paper to each student and have him write his name on it and put it into a small box which you provide. Then have students draw out another student's name. Throughout the rest of the day, students should watch for actions that they can appreciate about each other.

Share Inspiration

Have students make a "Thank You" card for the person whose name they drew. The card should include a written statement identifying the quality that is appreciated as well as a specific example of how or when the quality was expressed.

Skills for Living

Water Is Abundant

Objectives	• Identify various Skills for Living and recognize their use in the classroom
Benchmarks	• Contributes to the development of a supportive climate in groups (Life Skills: Working with Others 1, Level IV); Identifies personal strengths and weaknesses (Life Skills: Self-Regulation 2, Level IV)
Skills for Living	• All of those listed on page 41
Materials	• Assortment of magazines with pictures of people—several per student; rainforest tree drawn on butcher paper; leaf shapes cut out of green paper—1 per student; small box to hold leaves; scissors—1 per student; glue—enough for all students to share
	• Teacher Preparation: Draw the trunk and branches of a large rainforest tree onto a piece of butcher paper. Write one Skill for Living on each leaf and place it in a small box. Depending on the size of the class, some skills may be used more than once.

Awaken Enthusiasm

Tell students that the Skills for Living are abundant all around them. There are many examples of people demonstrating qualities that will lead them to happy and successful lives. Have each student draw a leaf out of the box. Tell students to work in pairs to discuss examples of their Skills for Living in action at home or school. For example, generosity might include sharing your lunch with a friend, letting someone ride your bike, loaning your brother some money, buying a birthday present for your cousin, bringing cookies to share with your whole soccer team, or donating some of your toys to a homeless shelter.

Focus Attention

Once pairs have an opportunity to discuss ideas, have each student read his Skill for Living to the rest of the class, give one example of the skill in action, and tape the leaf on one of the branches of the tree. An alternate or additional way to familiarize students with examples of the Skill for Living is to have students work in pairs to act out or pantomime a situation in which they use the skill.

Direct Experience

Have students notice that the leaves they have added begin to fill out the tree, but it needs "something more" to really be complete. That "something more" is pictures that illustrate each Skill for Living. Tell students that they are going to make a collage by cutting out pictures from magazines that show people (or animals) putting one of the Skills for Living into action. Make magazines, scissors, and glue available. As students cut out pictures they can glue them to the poster near the leaf that has the appropriate word. Tell them to do their best to have at least one picture for each Skill for Living. (If using this activity as part of the six-week unit, you may want to substitute making a collage with the following activity: Have each student draw a picture illustrating a Skill for Living and display the pictures around the room, with a small piece of paper taped below each picture. During the six-week unit, have students keep track of the skills being used in class by putting a "hatch mark" on the paper below the appropriate picture.

Challenge students to demonstrate each skill at some time during the unit so that each picture has at least one mark below it.)

Share Inspiration

Keep the poster displayed in a prominent place in the classroom. Acknowledge students when they are being an example of one of the Skills for Living in action.

Skills for Living

Sensitivity
showing understanding, caring, and kindness

Cooperation
getting along with others

Respect
showing regard for the worthiness of all creatures

Cheerfulness
having a positive attitude

Integrity
being truthful and honest

Concentration
being able to focus attention

Aspiration
striving for "personal best"

Self-Control
behaving appropriately

Vitality
being awake and ready

Perseverance
overcoming obstacles, determination

Servicefulness
helping others

Orderliness
being neat and organized

Dependability
being able to be counted on

Self-Reliance
taking responsibility for yourself

Creativity
approaching situations in fresh, new ways

Practicality
having common sense

Generosity
sharing with others

Introspection
being able to look within yourself

Problem-Solving
being solution-oriented

Curiosity
having a sense of wonder about the world

Appreciation
expressing gratitude

Maturity
the ability to relate to others' realities

The above list represents some of the skills which Education for Life uses in its classrooms.

Awareness

Water Is Everything

Objectives	• Increase awareness of the surrounding natural world
Benchmarks	• Knows that human beings can detect a tremendous range of visual and olfactory stimuli; understands that paying attention to any one input of information usually reduces the ability to attend to others at the same time (Behavioral Studies 3, Level III)
Skills for Living	• Awareness
Materials	• Paper and pencil

Awaken Enthusiasm

Begin by telling a story about a school of fish that lived at a tropical coral reef. They all enjoyed swimming around together and then one day one of the fish asked the others, "What is water?" All the fish looked at each other and no one really knew what water was, so they decided to go to a big, wise fish. They all swam over to the other side of the coral reef and asked the big fish, "What is water?" The big fish replied, "Water is above you and below you. It is to your left and your right. It is in front of you and behind you. It is within you and without you. You live all of your life in an ocean of water." The little fish all nodded and happily swam away, but the next day one of them looked at the others and asked, "What is water?" Tell the students that this story can be understood in many different ways. Ask them what they think it means. Tell students that one meaning from the story is that sometimes people don't notice or understand things that are around them all the time; they aren't aware. Explain that with practice everyone can become more aware and perceptive.

Focus Attention

Ask students to look carefully at the underwater and beach scene in *A Drop Around the World* for 30 seconds. (It is helpful if each student has his own copy of the book or small groups of students can cluster around a few books.) At the end of 30 seconds, close the books and ask the students several perception questions. Have them write their answers on a piece of paper so no one blurts out the answers. What color was the bucket the little child was playing with? (Red.) How many striped fish were in the picture? (Three.) What was the biggest boy holding? (A watering can.) Describe the flag. (Blue background with white stars and a red and white cross with a red and white "X" behind the cross.) Was there an octopus in the picture? (No, there was a red squid.)

Direct Experience

Tell students that they are "warmed up" and ready to play a game called "I Am Aware Of." Divide the class into small groups of eight to ten students each. Go outside into a natural environment and spread the groups out so they cannot easily see each other. Have each group form a circle. Tell them that they are going to finish the sentence "I am aware of ...." by saying a word or phrase about something they see, hear, smell, or feel. For example, "I am aware of the sunlight sparkling on the pine needles of the tree." "I am aware of the wind blowing across the grass." "I am aware of how hot the sun is on my shoulders." Encourage them to increase their awareness by using all of their senses. Each student should take a turn, going around the circle. After a student shares his awareness, the others should also try to become aware of what he just mentioned. Talking between students should be minimal. Continue for several times around the circle.

Share Inspiration

Bring all of the students back together and have them share any awarenesses that they think are especially unusual or unique from their group. Students could write a haiku as a follow up activity. A haiku is a type of Japanese poem that consists of three lines. The first line has five syllables, the second line has seven syllables, and the third line has five syllables. The following three haiku poems were written by sixth grade students:

A Feather in a Web

A feather quivers
As it settles in a web,
Shimmering with dew.

After a Rain Storm

Rain on a pine tree,
Sparkling and shining like stars.
It is beautiful.

Fall

Orange, red, golden
Flittering in the wind,
a Maple drops its leaves.

Calmness

Water Is Frozen

Objectives	• Practice a stress-reduction technique to become still and calm
Benchmarks	• Knows strategies to manage stress and feelings caused by disappointment, separation, or loss (Health 4, Level III)
Skills for Living	• Calmness
Materials	• Tape of lively music

Awaken Enthusiasm

Explain to students that because water in the polar regions is frozen the class will play a game of "Freeze." Put on lively music and instruct students to move around the room to the music. They can be fluid and flowing like water that is liquid. Tell the students that the music will periodically stop. When it stops they should freeze, like the water at the Arctic or Antarctica. Identify any students who move once the music is stopped. Tell them that they have "thawed" and have them sit down at their desks. Continue playing until most of the students are seated.

Focus Attention

Once all students are seated, challenge them to sit perfectly still and to not move for two minutes. Tell them that they can blink and breathe, but not swallow or move. Set a timer for two minutes. Ask students how well they did at sitting still.

Direct Experience

Explain to students that one way to become still and calm is to breathe rhythmically. Have students sit up straight at their desks and tell them to inhale to a count of six, hold their breath to a count of six, and exhale to a count of six. Count out the rhythm for several rounds of breathing.

Share Inspiration

Tell students that they will have a chance to meet the stillness challenge again, using the breathing technique. Set the timer for two minutes and have students practice breathing and sitting still. When finished, discuss with the class if they could tell a difference between their first try and their second try of sitting still. Ask students what thoughts or images came to their minds that helped them. Explain that breathing rhythmically while sitting still is a good way to relieve stress and to help them to calm down after being upset.

Responsibility

*Water Is
Used*

Objectives	• Practice water conservation at home and school
Benchmarks	• Understands how human actions modify the physical environment (Geography 14, Levels I—IV)
Skills for Living	• Responsibility
Materials	• Poster board or butcher paper for water conservation chart; plastic gallon jug filled with water

Awaken Enthusiasm

Explain to the class that communities are formed whenever people live together in one place, and can generally be divided into three classes— villages, small towns (including suburbs of large cities), and cities. Cities are the largest communities where people live and work. Although cities are not a natural region of the world, they have a dramatic impact on the environment. Large populations of people use vast amounts of natural resources, including water. Divide the class into small groups and ask each group to brainstorm all of the ways they use water, either directly or indirectly. Encourage them to remember less obvious ways such as growing their food.

Focus Attention

Compile the different group responses to create a class list. Tell students some interesting water facts (use a plastic gallon jug filled with water as a visual aid to illustrate the amount and weight of one gallon of water): about 75% of the water used in the home is used in the bathroom; each American uses about 80 gallons of water a day; it takes 1,630,000 gallons of water to feed an American for a year; it takes 100 gallons of water to produce one pat of butter; only 3% of the Earth's water is fresh water; the US. uses 450 billion gallons of water every day. (These facts are taken from the book *50 Simple Things You Can Do To Save The Earth* by the Earthworks Group ©1989.)

Direct Experience

Explain to students that it is important to use water responsibly. Ask them how they think water might be conserved. Compile their suggestions on a piece of poster board or butcher paper. You may also want to incorporate the following suggestions and information:

- Turn the water off when you're brushing your teeth; just wet and rinse the toothbrush. (Saves 9 gallons of water each time you brush.)
- When washing dishes by hand, fill up a sink with rinse water rather than leaving the water running. (Save about 25 gallons.)
- Put a plastic bottle in the toilet tank to reduce the amount of water used for flushing. Be sure to soak off the label first, weight it with stones, and fill it with water. Place it in the back of the toilet tank in such a way that it doesn't interfere with the flushing mechanism. (Save 1—2 gallons per flush.)
- Take a shower instead of a bath. (Save about 25 gallons.)
- Don't let the water run when you want a cold glass of water. Keep a container of water in the refrigerator if you want it cold. (Save 3 gallons.)
- Organize a group, such as your class, to adopt a stream, pond or beach. Patrol the banks or beaches picking up trash and litter. (Save the water from being polluted and from being an eyesore.)

45

- Ask your parents to install a low-flow shower head. (Save 2.5 gallons per minute.)
- Ask your parents and your principal to install low-flow aerators on all water faucets (save up to 2 gallons a minute) and to use unbleached paper goods such as paper towels and toilet paper. (Save streams and rivers from becoming polluted.)
- Ask your parents to use low-phosphate detergent. (Save lakes and streams from dying.)

Share Inspiration

Ask students to identify the conservation measures they can follow during the next week. Tell students to write their name or put a check mark on the chart each time they do one of the suggestions to save water. At the end of each day or at the end of the week, calculate how much water was saved. Extension: Cities vary greatly around the world and can be studied by country, continent, or culture. Ask students to research major cities of the world, identifying the natural habitats that are most directly affected by each city's existence.

The Water Cycle

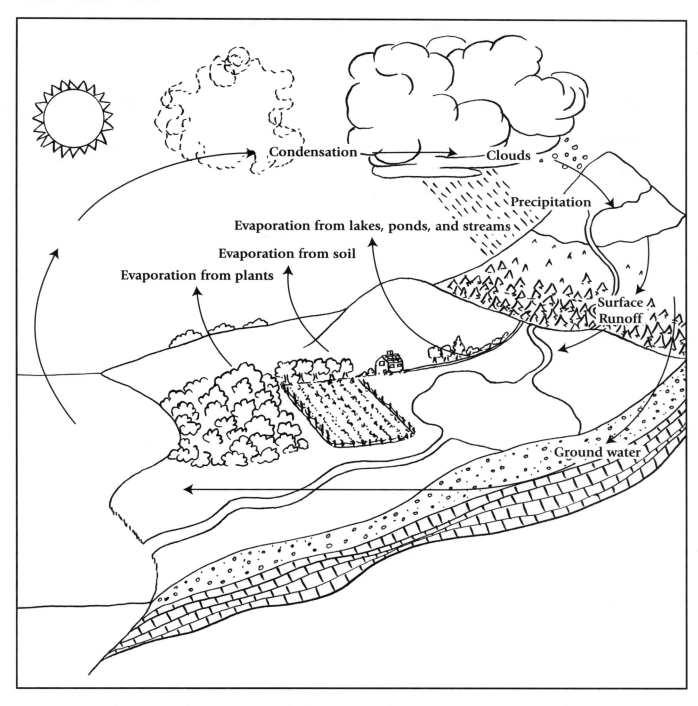

Copy Master (see Transformation Tricks lesson plan)

Resources

Smart Moves: Why Learning Is Not All In Your Head, by Carla Hannaford
Your Child's Growing Mind: A Guide to Learning from Birth to Adolescence, by Jane Healy, Ph.D.
Mapping Inner Space: Learning and Teaching Mind Mapping, by Nancy Margulies
Integrated Thematic Instruction: The Model, by Susan Kovalic
Super Teaching, by Eric Jensen
Training the Teacher as a Champion, by Joseph Hasenstab
Education for Life, by J. Donald Walters
Tribes: A Process for Social Development and Cooperative Learning, by Jeanne Gibbs
Megaskills: How Families Can Help Children to Succeed in School and Beyond, by Dorothy Rich
Self-Starter Kit for Independent Study, by E. Doherty and L. Evans
Ranger Rick's Nature Scopes: Tropical Rainforests, Diving into Oceans, Discovering Deserts, and
    Wild About Weather, by The National Wildlife Federation
Brain Gym, by Paul and Gail Dennison of the EduK Foundation
Habitat music tape, by Bill Oliver, Glen Waldeck and the Otter Space Band (1-800-95-OTTER)

Related Resources from Dawn Publications

A Walk in the Rainforest, by Kristin Joy Pratt
A Swim Through the Sea, by Kristin Joy Pratt
Sharing Nature with Children, by Joseph Cornell
Sharing the Joy of Nature, by Joseph Cornell
This Is the Sea That Feeds Us, by Robert F. Baldwin
Wonderful Nature, Wonderful You, by Karin Ireland

DAWN Publications is dedicated to inspiring in children a sense of appreciation for all life on Earth. For a copy of our catalog, or for information about school visits by our authors and illustrators, please call 800-545-7475. Please also visit our web site at www.DawnPub.com, or e-mail us at nature@dawnpub.com